Idols of the Mind

vs.

True Reality

Bhakti Madhava Puri, Ph.D.

Readers interested in the subject matter discussed in this book are encouraged to contact:

B. Madhava Puri
princeton@bviscs.org
www.bviscs.org

Copyright ©2020 by Bhakti Vedanta Institute of Spiritual Culture and Science

All rights reserved. This book contains material protected under International and Federal Copyright Laws and Treaties. Any unauthorized reprint or use of this material is prohibited. No part of this book may be reproduced or transmitted in any form or by any means, electronic or mechanical, including photocopying, recording, or by any information storage and retrieval system without express written permission from the author / publisher.

Published by Bhakti Vedanta Institute of Spiritual Culture and Science
Printed in the United States of America

Cataloging-in-Publication Data

Puri, Bhakti Madhava
      Idols of the mind vs. true reality /
by Bhakti Madhava Puri
         Includes bibliographical references
         ISBN: 978-1-7349089-5-4

Library of Congress Control Number: 2020909326

Hegel's idea is that realism is there. Ideal realism. The reality is an idea. It's the beginning of everything; the idea first. The idea is not an abstract imaginary thing, but it is real. Everything is the effect of the idea, starting with the idea and then it is translated into action. Everywhere we find that. Ideal realism - that is Hegel's theory, and ours of course - consciousness first. Then, these are all the effects of consciousness.

-Srila Bhakti Rakshak Sridhara

Foreword

In the era of IT and robotics it is all too easy to confuse digital artifacts with living systems but, as Madhava Puri clearly shows in this wide-ranging essay, the two are irreconcilable. A living organism is not a machine. It is a construct of human ingenuity and relies on external mediation to function. Yet practitioners in the field prefer to imbue their constructs with metaphysical attributes, ascribing intelligence to problem-solving software, and life to robotics. The public are more easily seduced by this delusion that many might think. On the Jimmy Fallon television show a mechanical mannequin was produced. The host (and the guest who had invented it) agreed it was basically alive. It was, of course, no more than an elaborate doll; but the *zeitgeist* has robots as near-rivals to humankind. A popular television show reflects what is current in the minds of many, and this serves as a warning to us all.

When so many authorities are promulgating their products as almost supernatural, the Bhakti Vedanta Institute of Princeton seeks to investigate the realities of this intricate interaction to provide a soundly-argued case that restores reality. With its roots in Aristotelian philosophy, the distinction between life and non-life has an ancient lineage. Yet in our post-Cartesian mechanistic environment it is all too easy to subsume the one into the other. Seventy years ago, at Bristol in England, Grey Walter invented robots the size of a helmet that were equipped with photo-electric sensors which caused them to avoid dark obstacles placed in their path. They moved like small cybernetic entities, and he went so far as to insist that he had created the first 'artificial life', with the equivalence of two living cells. Indeed, he gave them a Linnean binomial designation – he said they were *Machina speculatrix*, a new organism. Throughout the following decades, this hubris has marked out our progress through to a digital age.

To those of us who are devoted to the elucidation of the single cell, and the adaptability, ingenuity, and sheer intelligence it manifests, the pretense of making mechanical models is absurd. There is a majesty about the vitalism of a microbe, with which technological models cannot compare. A machine is just a machine, no matter how refined we make it appear, and the relationship between organism and artifact becomes clearer as the reader digests the arguments within these pages.

In this book, Madhava Puri seeks to transcend the physical and seek enlightenment from the spiritual dialectic of G. W. F. Hegel as well as the Bhagavat Vedanta of Indian philosophy. How well he succeeds is up to the reader to determine. The fundamental thrust of the discussion is to enlighten us all as to the down-to-earth, pragmatic realities of being. We explore. We sense. We think. We conclude. We live. A machine merely exists, and comes to being only through the artifice of its manufacturer. Complex devices, like autonomous robots, can behave in extraordinary ways and perform unimaginable tasks, that's true. But our sense of wonder and amazement is wrongly directed if we find we are admiring the machine. This sophistication reveals the mental might of the man who made it, and is testimony to the complexity of life. The machine would be nothing without its human creator, and we would do well to remind ourselves where our loyalties lie.

The title of this book, *Idols of the Mind*, stems from the writing of Francis Bacon in the sixteenth century. It was at this time that Juanelo Turriano, a clockmaker to the Holy Roman Emperor Charles V, constructed a mechanically operated model monk which walked and prayed silently as it brandished its crucifix in one hand, and a rosary in the other. Charles's son, King Philip II, commissioned the construction of this curious automaton and, when it was set in motion, courtiers fled in fear, believing it to be alive. We now smile patronizingly at their naïveté, knowing better in the twenty-first century. Or do we? Madhava Puri demonstrates that we, in our own era, are just as confused, and are equally easily taken in.

Even the most sophisticated digital device is no match for the wonderful workings of a microbial cell. Life has a degree of refinement that cannot thus be replicated. As we explore the interrelationships between humanity, the mind, consciousness and the constructed habitat in which we exist, we would do well to appreciate the uniqueness of living, and the arbitrariness of existence. No mind can apply itself to comprehend anything as great as it is, and the chance to explore the ramifications of this enduring paradox within the pages of this book is one in which we can relish – as we celebrate our uniqueness in an unfathomable world.

Brian J. Ford
Cambridge, England

Contents

Preface..3

Logic of Life...11
   Mechanical Objects   16
   Chemical Objects   17
   Biological Systems   18
   The Objectivity of the Organism   22
   Conclusion   23
   References   25

Idols of the Mind vs. True Reality..............27
   Uncertainty and Unknowing   27
   Explanation and Correspondence   30
   God, the Universal I is also Self-consciousness
      and not merely Consciousness   36
   Free Will and the Fall   40
   I and Mine   41
   Reality Has Its Own Purpose In and For Itself 47
   References   53

**The Unreasonable Effectiveness of
Mathematics in the Natural Sciences**..........57
   The Identity of Indiscernibles   57
   Galileo's Error of Primary and
      Secondary Properties   58
   Philosophical Development of Ideas
      that Science Ignores   59
   From Pantokrator to Chaos   61
   Logic of Mathematics Compared
      to the Logic of Nature   62

References 65

The False Elephant and the False Ego............69

Unity of Science and Religion.....................71
   The Link Between Science and Religion   71
   Importance of Concepts and Their Content   72
   Knower: Knowledge: Known   74
   Four Aspects or Types of Cause   76
   Descartes: Subject-Object Duality
     of Consciousness   78
   Levels of Consciousness   79
   Science Has Lost Its Self-consciousness   81
   Isolated and Independent False Ego   83
   The Principle of Bhakti   84
   References   85

History of the Princeton
Bhakti Vedanta Institute........................87
   References   91

Contact Us...................................93

Preface

Francis Bacon (1561-1626) placed the idea of "Idols of the Mind" at the root of what would become modern science. He had a clear understanding of the limitations of the finite human mind and intellect, and that any endeavor for scientific knowledge must begin with acknowledging that limitation.

In his *Instauratio Magna* (1605), Bacon wrote:

> "The mind, hastily and without choice, imbibes and treasures up the first notices of things, from whence all the rest proceed, errors must forever prevail, and remain uncorrected. . . .
>
> The human understanding when it has once adopted an opinion (either as being the received opinion or as being agreeable to itself) draws all things else to support and agree with it. And though there be a greater number and weight of instances to be found on the other side, yet these it either neglects and despises, or else by some distinction sets aside and rejects, in order that by this great and pernicious predetermination the authority of its former conclusions may remain inviolate."

He defined Idols of the Mind as the numerous varieties of errors in mental processing. In this case, an idol refers to an idea that is held in the mind which receives veneration but is without substantial truth in itself. It refers more to a mental fixation than a symbol or statue. We might refer to it as a model, and even extend that to include the idea of paradigm or paradigm shift, which Thomas Kuhn (1922-1998) introduced in his book *The Structure of Scientific Revolutions* (1962). One of the articles in this collection is concerned with this important topic which also appears in the title of this publication.

The Indic conception of *Maya* plays a very important role in the Bhagavad-gītā when it comes to understanding the true nature of the self beyond the material misconception of the world. This Sanskrit word consists of two parts, *ma-ya*, which means "not" (*ma*) and "that" (*ya*). It refers to considering something to be what it is not - an illusion, like mistaking a rope to be a snake, or the appearing phenomenal world to be what Kant called the noumenal reality.

A scientific study of wrong conceptions is just as important as developing the right ones, and this seems to have been recognized by the ancient as well as more recent traditions of knowledge to which we are heirs. For this reason, it is important to take a closer look at some of the modern ideas we uncritically accept as certain and more closely examine some of the true or mistaken conceptions they may embody.

Some of the inquiries we can make are: Does Life come from Matter, or Does Life come from Life? Darwin's theory of evolution, or the neo-Darwinian synthesis, is based on purely mechanical or chemical conceptions of living organisms, including ourselves. However, we now understand that consciousness plays a crucial role in the functioning and survival of organisms at all levels. Darwin's theory thus turns out to be a theory of a world that never existed because Nature does consist of conscious living entities. If we were to analyze the patterns of traffic based only on the mechanical movement of inanimate cars without recognizing the essential role of cognizant drivers, we would not expect a proper understanding to result from such a study.

In his recent book, *The Case Against Reality: Why Evolution Hid the Truth from Our Eyes* (2019), Donald Hoffman, a cognitive scientist, cogently argues that if evolution were true it would have hampered Man's fitness to properly comprehend reality, because "natural selection has favored perception that hides the truth..."

Biologists today do not know the minimal number of chemicals needed to form a living organism. Yet scientists studying the origin of life propose that a sequence of fortuitous steps (chemical reactions) randomly occurred to produce the first biological life. Certain steps in that sequence may have happened randomly, but to claim that an unspecified large number of steps in that sequence occurred fortuitously only adds to the rational incredulity of such an improbable scenario. As George Wald said, a scientist may have to "choose to believe in that which I know is scientifically impossible, spontaneous generation leading to evolution." [1]

One of the most important topics to consider is the role of mathematics in science, and if it is reasonable or unreasonable to expect the applicability of abstract mathematics to the scientific understanding of a substantial and living Nature. This is briefly considered in one of the chapters of this book.

The Princeton Bhakti Vedanta Institute is concerned with the study of these and other topics. We are involved in the study of the role of consciousness in science, as well as the subjective evolution of consciousness and its higher development, especially as found in the philosophical schools of Yoga and Vedānta. Our method is based on the central role of the Conceptual Realism interpretation of the Hegelian *Science of Logic* and his *Encyclopedia*, which is currently undergoing a renaissance of academic interest and new insights.

The modern scientific focus on mechanistic reductionism studies Nature in its isolated atomic or molecular parts, whereas the systems of Nature exhibit that of Organic Wholes. Each part thus has two aspects, its being-for-itself or its atomic features, as well as its being-for-other or its relational features. Considered as a positive whole both features must be accounted for, as a negative unity the Whole sublates (retains yet negates) its parts.

Furthermore, modern science develops its theories/ concepts of Nature as if consciousness, the cognitive/ conceptual side of empirical perception does not contribute to the data of observation when in fact it is essential to it. Science has left out the role of the scientist in its picture of Nature. This results in or arises from the dualist conception of reality in which consciousness is isolated from its objects or content.

An additional isolationist/divisional principle of the modern scientific view is that consciousness is isolated within each individual ("I"). Collective consciousness ("We") or universal consciousness is not considered. Individuals are conceived as atomic sentient entities in/for themselves exclusively in the modern viewpoint. But simultaneously they may also be considered organic parts, moments, or instantiations of the universal conscious being of the Organic Whole. In a more comprehensive sense, the parts are negated in and as the Whole since they lose their independent being therein.

Vedānta philosophy is established on a more holistic ontological and epistemological basis from its inception. Indologists have discovered the universal principle of Bhakti that systematically unites the main divisions of Vedic philosophy (*karma, jnana, bhakti, yoga*). [2] Current scholars in the field have shown how this principle is logically intrinsic to the Plotinian conception of the One. [3, 4] The Hegelian concept of being-for-other, and especially his idea of the Absolute Truth as being-in-and-for-itself comprises not only the relation of the parts to the whole but in its negative sense to the self Concept. Thus the universal principle of bhakti (dedication/devotion) to the Absolute can be rationally understood as a logically scientific one. Thus it has universal scientific, social, moral, ethical, religious, and environmental significance that may be developed broadly and comprehensively.

The internal/intrinsic unitive principle of bhakti is called *rasa*, which we may translate as "taste." This is a wider, more comprehensive topic that takes us deeper into the Vaishnava tradition of India that transforms bhakti itself into its highest, most ecstatic form - divine love, *prema*.

This work and the mission of the Institute have been inspired by the grace and mercy of the rich and extensive philosophical and spiritual culture exemplified and embodied in the teachings of our Brahma-Gaudiya Vaisnava predecessors, *guru varga*. We offer our humble and loving acknowledgment to them while accepting all shortcomings and misinterpretations as my own.

-B Madhava Puri

References

[1] Wald, G. (1978). Origin, Life and Evolution. *Scientific American*.

[2] Biardeau, M. (1989). *Hinduism: the Anthropology of a Civilization*. New Delhi: Oxford University Press.

[3] Butler, E. P. (2018). Bhakti and Henadology. *Journal of Dharma Studies*, *1*(1), 147–161. doi: 10.1007/s42240-018-0004-6

[4] Adluri, V. (n.d.). Philosophical Aspects of Bhakti in the Narayaniya. Retrieved from www.presocratics.org/wp-content/uploads/2015/06/Vishwa_Adluri_Philosophical_Aspects_of_Bhakti_in_the_Narayaniya_WSC.pdf

Logic of Life

Modern science generally assumes that the same laws of logic apply to mechanical, chemical and biological entities alike because they are all ultimately material objects. This may seem to be so obvious that there would be no need to validate it - experimentally or logically.

In this section, we would like to critically examine this assumption and show that from an empirically observational level, as well as from a rational/logical level, it is not valid. This becomes apparent, for instance, when we consider the simple observation in which we distinguish animate from inanimate objects: those objects that seem to spontaneously move by themselves and those that move only when impelled by some applied force outside or beyond the object. This distinction may be valid at the macroscopic level more than at the level of theoretical atomic particles. Thus the detailed nature of spontaneous movement must also be understood.

We consider animate objects to be living, and the inanimate ones dead. Yet we consider both as being material objects since they are both composed of atoms and molecules. Even if the composition may be a little different for the two, still the living objects can die and thus become the same as the dead objects. Thus the difference does not seem to be specifiable within the material aspect of the object.

This means we are left with the question: What is it that automates animate living objects that seems to be missing in dead ones? At first, we may try to answer this question by claiming that it is the chemical reactions going on in the animate object that are causing it to move. After all, chemical reactions can occur on their own in any laboratory by a process as simple as mixing two reactive chemicals.

Of course, there is a serious problem with that explanation. Chemical reactions generally produce a stable product - just like acid and alkali when added together produce a salt. The reaction seems to occur spontaneously, but it does not go on and on for many years. It does not sustain itself. In the living object or living organism there occurs sustainable chemical activity of a special type called biological activity. That activity can become very complex, often defying all explanations at a simple chemical level because of the intricacy involved.

But living organisms exhibit further peculiar traits, which we call behavioral symptoms, that are not found in inanimate objects. Organisms exhibit growth, irritability, reproduction, metabolism, etc. The point is that animate and inanimate objects, even at the simple level of casual observation do exhibit important differences. Objects that participate in chemical reactions are different from mechanical objects that do not transform with each other. And animate objects or organisms behave in manners that chemical objects do not exhibit.

Over two millennia ago, Aristotle attempted to explain the peculiar nature of living organisms by philosophical analysis. He called dead matter *dunamis*, or potentiality, and matter in action *energia*, or actuality. The word actuality implies act. The Greek word *energia* means energy. According to scientists, energy means "the ability to perform work." A certain amount of energy can do a certain amount of work. This is how energy is determined and measured.

Yet, what is it that moves dead matter (*dunamis* or potentiality) into action (*energia*, act-uality)? Aristotle called that actualizing force *entelecheia* or entelechy. This peculiar word comes from *teleos* or teleology, and specifically refers to inner (*en*) teleology. It means intrinsic/natural purpose or end in the sense of aim.

We may at first think of teleology as the external purpose, as is the case when a carpenter builds a chair from wood. The wood is the original matter, and the chair is the end product - the idea of the chair in the carpenter's mind is actualized in the form of the wood as a chair. Any artifact can be viewed from this perspective of external teleology.

But inner teleology is quite different. According to Aristotle, there are many types or kinds of being or (though strictly not equivalent) matter. For instance, the being of an animal is different from the being of a bird or of a man. Thus the *dunamis* or potentiality has different potencies depending on the kind or species of creature that it is. For example, the seed of an oak tree, or the egg of a chicken has certain potencies within them characterizing the type of matter they are. When their potency is awakened their entelechy will drive them to actualize as a tree or chick.

If we consider Aristotle's ideas from the modern viewpoint, we find a similar concept is utilized in biology. The specific genetic and phylogenetic material of each type of organism is unique due to the different arrangements of the amino acids in the DNA code and other specific proteins, enzymes, mitochondria, etc. that are part of the makeup of the various kinds or species of life. The specific type of matter will, therefore, determine what kind of creature will develop from it. So it is a tribute to the brilliance of Aristotle that his conception, in principle, is still quite valid even today.

Of course, modern science has not discovered what corresponds to the entelechy, the mysterious force that causes a particular glob of protoplasm to differentiate and almost magically develop into whatever life form it eventually becomes. It is much too complex and specific to be understood as the result of a series of standard chemical reactions. Various experiments have been done on the zygote (fertilized egg) to show that there is a definite directive process involved that continues despite severe modification of the basic structure at an early stage of development. [1]

It is almost as if there were an invisible pattern, concept, or idea that was imprinted in the specific type of matter that directed it toward development into the specific creature that it becomes. Aristotle considered the situation from this point of view and concluded that there is a soul that was responsible for this. [2] A couple of thousand years later, G.W. F. Hegel also demonstrated in his *Science of Logic* [3] that there is a Concept involved in the determination of its corresponding content. In between these two towering figures of philosophy, Immanuel Kant also developed a similar theme (he called *Naturzweck*) in his philosophical analysis of the scientific understanding of organisms. [4]

I think it will be very useful to look at the way Hegel organized the various types of objects that we observe in Nature, viz. the mechanical, chemical and biological, according to what he called the Concept (*Begriff*). The Concept, for Hegel, is essentially a dynamic or organic unity of the different moments or parts that constitute its content. We will start with his application of this idea to the mechanical object.

Mechanical Objects

Mechanical objects do not have an internal relationship of parts. Thus you can divide a rock and it becomes two rocks, but the basic nature of the rock does not change. What lacks internal relation like this, is said to have merely external relation to what is other than itself. Thus rocks are related to other rocks by the external force of gravity, or other causal factors. Objects that lack internal relatedness may possess merely external relatedness. Planets relate to each other externally, as in the solar system, explicable by the laws of gravity and motion. Newtonian gravity depends upon mass, but the internal composition of that mass does not play any role in determining their attraction to other planets. Thus gravity acts in a purely external way to unite the planets as a solar system moving around the Sun.

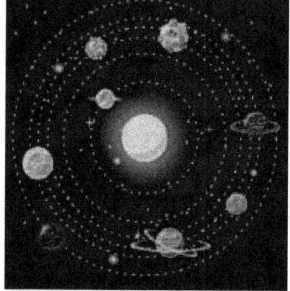

In mechanistic objects, the unifying Concept (in this case, gravitational force) exists only implicitly (as a principle), and therefore only explicitly or externally to the objects. Mechanics views a system as having separable, independent parts that possess a fixed identity outside their connection within the system of which they are parts. If the isolated parts of a system retain the same identity as when connected within it, it is called a mechanical system. This is the particular logical character or nature that is implied when we refer to a system as being mechanical.

Chemical Objects

Now, those entities that show an intrinsic affinity (chemical affinity) toward other entities leads to the next type of object - the chemical object.

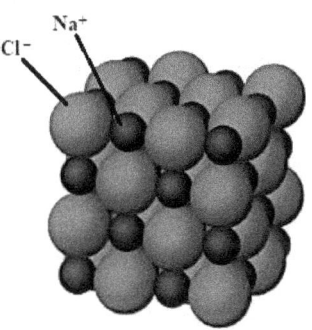

Chemical objects have parts that are internally related. They are not the same when isolated from each other as when they are connected or united with each other. Thus, for example, a salt crystal cannot maintain its identity when divided at its most fundamental molecular level since sodium and chloride atoms when divided would form two distinct substances - sodium and chlorine. External relations are formed due to the intrinsic properties of the individual parts of a chemical reaction. Thus an acid is intrinsically related to an alkali, which combines to form a neutral salt. Their unity, the neutral salt, is a completely different substance compared to the distinct parts in their isolation.

Furthermore, to speak of nascent acid would be a misnomer. A substance is acidic only in relation to alkaline substances. Its identity or definition as an isolated entity is incomplete and can only be understood in its relation with another object. On the other hand, mechanical objects possess an individuality that is complete in itself without reference to another object.

The unity of a mechanical system, like the solar system, made up of mechanical objects, is established externally in the form of a law, which reigns outside of and over the parts and by which the parts of the system are regulated. On the other hand, the unity of the chemical system is intrinsic to the parts, arising from their intrinsic natures. The ordered structure of a crystal is based on the nature of the constituent parts of a chemical system. Still, the parts of a chemical system retain their identity even apart from the interactive system, so that their initial and final states can be differentiated. In this sense, the parts are both independent as well as dependent. For example, acid and alkali can be isolated in different bottles and then added together to form a third substance – a neutral salt.

Biological Systems

Those parts that can not be separated from a system without destroying it as a working system, can no longer be called parts but are participants or members of a dynamic whole. The participants are as essential to the whole as the whole is to the participants - this is the biological system or organism. Here we are removed from the stasis of fixed objects and are in the milieu of pure dynamical activity. Participants cannot be isolated from the whole in which they are participants and remain what they are. A DNA molecule can no more be what it is as a producer of protein molecules than the protein molecules can be what they are as produced from the action of DNA, and producing the DNA. Each

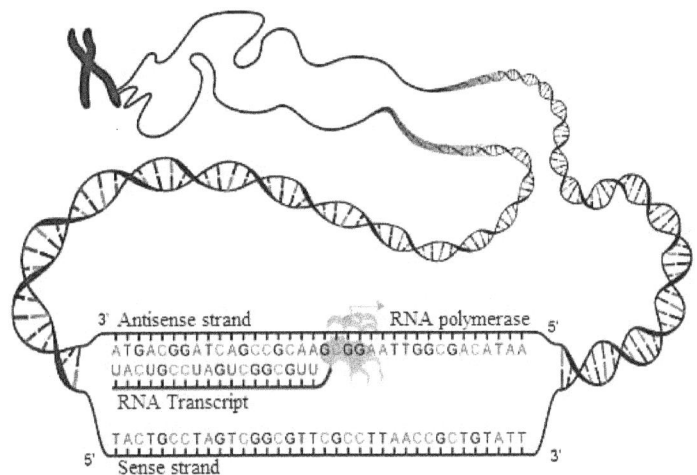

participant is the cause and effect of every other participant, as Kant defined organism. Therefore nothing in an organism is without purpose, nor is the organism as a whole without purpose in the environment. Thus everything in the organism is both purpose (end) and means.

Life is a unity in multiplicity. It is a process as a united flow, but it consists of many instantaneous moments - like the frames of a movie. Abstract understanding tends to think of either a unity or a multiplicity. Pure multiplicity is indicative of the atomic thinking of material reductionism. Pure unity is the indeterminateness of abstract monism. Unity in multiplicity is the comprehensive thinking of dialectical reason. Life has to be comprehended as a process in which its participants are simultaneously both ends (products) and means (production) to one another.

The living organism internally assimilates itself and produces itself. This self-consumption and self-production is its metabolism, by which it anabolizes and catabolizes, creates and destroys its own cellular substructure to maintain its own superstructural integrity. Likewise, the superstructural system assimilates the outer environment of which it is an integral participant. It both consumes and produces the environment in which it lives, but on a localized scale, unlike the totality of its destructive and constructive activity that occurs within itself. Reproduction is a production of itself as a totality but in a localized portion of the environment. It is a process of preservation of the species.

Its inner metabolic process is the preservation of its particularity. Its assimilation and defense against the environment is the preservation of its individuality. And the reproduction of itself as a species is the preservation of its universality. The particular, individual and universal aspects of the living process are characteristic of what is called a Concept. A Concept has three aspects: universal, particular and individual. For example, in biology, we speak of genus, species and specimen. Mammal is a genus (general or universal), whereas tiger is a particular species or kind of mammal, and the individual tiger that we meet in the jungle is a specimen. All three of these aspects are required to completely specify the individual identity of whatever is experienced.

The living organism, therefore, ultimately has the Concept as its substance when it is comprehended completely. Thus, the categories of understanding that seek to fix identities in their identity (e.g. A=A, B=B, etc.) do not apply to the living organism whose participants do not possess isolated identities but are identified only in their mutual relations. Unlike chemical objects, the participants of a biological system are produced by, as well as productive of, the other participants.

The proper understanding of inner teleology requires that we grasp that there is not one thing being driven by another outside of it or beyond it, but a single nature actualizing itself, sustaining its own reality. Teleology is in the organism in the same way that reason is in the thing studied. The self-differentiating unity of organic teleology is not observationally but conceptually grasped. In other words, in the same way that gravity can not be directly observed but is deduced from the behavior of bodies, we cannot observe teleology but it must be logically concluded because the behavior of the participants cannot be explained by either mechanical or chemical principles. As previously explained, that which is logically concluded involves thought, and thought when developed in form is called a Concept.

The Objectivity of the Organism

Hegel briefly summarizes what has been explained above in his *Science of Logic:*

In the first stage of comprehending the objectivity of the organism, when the Concept is not explicitly known and is thus only implicit or potential for knowledge as inner unity, we determine only the purely external relationship of parts known as the mechanism. Here, the totality of the determinations of the Concept appear merely as the external immediacy of its self-subsistent, independent parts, in other words, as an ordered aggregate.

In the second stage of comprehending the organism as an object, the immanent law of the parts is established so that particular relationships between the parts is revealed. This is chemism.

In the third stage, the essential unity of the object is comprehended as distinct from the self-dependence of the parts and posited as a subjective end which is opposed to the objectivity that it utilizes as a means to fulfill its purpose. This is teleology or the biological object.

This end or purpose is the Concept that is related to objectivity for the purpose of removing its defect as being merely subjective. As actualized end, it is the return of the Concept to itself from its externally posited being and in this internal unity with itself is called the Idea.

Conclusion

In his book, *This is Biology*, [5] leading biologist Ernst Mayr wrote,
> "It is a little difficult to understand why the machine concept of organisms could have had such long-lasting popularity. After all, no machine has ever built itself, replicated itself, programmed itself, or been able to procure its own energy. The similarity between an organism and a machine is exceedingly superficial."

Immanuel Kant, like Aristotle before him and Hegel after him, understood that an organism had to be distinct from both mechanical and chemical systems, and could only be understood within a teleological framework. For Kant, teleology exists when two criteria are met:

1. The parts of a whole are possible only through their relation to the whole.

2. The parts are combined into a whole by being reciprocally the cause and effect of their form.

He, therefore, claimed that "There will never be a Newton of a blade of grass." [6] This is because there is no regulative law that can be formulated for a teleological system. As previously explained, law applies only externally to mechanical systems, whereas teleology is an effect that is internal to the unity of the system.

What Hegel called the Concept, Aristotle called the soul. This additional element is needed to describe the living organism, and it cannot be completely explained without it. This is the conclusion of some of the greatest philosophers of Western culture. It has its counterpart in Eastern philosophy as well. In fact, it is the teaching of all the great religions of the world.

Only modern science has insisted on trying to explain life on a purely mechanical-chemical level, and has failed repeatedly to even come up with a definition of life on that basis, as it must since life and matter are inherently understood as being distinct principles. Reason is one, thus modern science, as the honest study of reality, must eventually concur with the same truths that human reason has established in our philosophical and spiritual traditions. It is due to the progress of science that we are led to acknowledge the limits of science and the importance of recognizing life as a distinct principle beyond the mere material or naturalistic conception of Nature. The whole concept of living Nature, itself, cannot be encompassed simply in terms of atoms, molecules, and their physical and chemical reactions. Deeper truths have to be sought in the corresponding reality of thought and spirit. It is hoped that this brief introduction to the logic of life will inspire further study into this deeper reality.

References

[1] Starr, C. (2005). *Biology: Concepts and Applications* (p. 650). Thomson Brooks/Cole.

[2] Aristotle. (1991). *De Anima (On the Soul)*. (R.D. Hicks, Trans.). Prometheus Books.

[3] Hegel, G.W.F. (1969). *Science of Logic* (p. 710) (A.V. Miller, Trans.). George Allen and Unwin, Ltd.

[4] Kant, I. (1766). *Dreams of a Spirit-Seer*. Forgotten Books.

[5] Mayr, Ernst, *This is Biology: the Science of the Living World*, Belknap Press of Harvard University Press, Cambridge, 1997.

[6] Kant, I. (1790). *Critique of Judgement*. (§§ 75, p. 282-283) (W. Pluhar, Trans.). Indianapolis, Indiana: Hackett.

Idols of the Mind vs. True Reality

Reason, Uncertainty. and Unknowing

> "[Isaac] Newton's goal was incomparably more vast than the discovery of the 'mathematical principles of natural philosophy.' Newton wished to penetrate to the divine principles beyond the veil of nature, and beyond the veils of human record and received revelation as well. His goal was the knowledge of God, and for achieving that goal he marshaled the evidence from every source available to him: mathematics, experiment, observation, reason, revelation, historical record, myth, the tattered remnants of ancient wisdom." [1]

It was from the wide breadth of his learning, yet single minded focus to comprehend True Reality that led Newton to humbly remark:

> "I do not know what I may appear to the world; but to myself I seem to have been only like a boy playing on the seashore, and diverting myself in now and then finding a smoother pebble or a prettier shell than

ordinary, whilst the great ocean of truth lay all undiscovered before me." [2]

In mathematics, the N-body problem for $N>2$ bodies interacting according to an inverse square law, was well known to Newton. This played a great role in his conception of the order and stability of the solar system. The later development of chaos theory by Poincaré and others recognizes this problem, as well as perturbations, and initial condition errors that are fundamental to computer-simulated stability calculations over reiterations of billions of years. [3] Through these methods, it has been found that ejections and collisions are possible within 5 billion years. Newton's prescient uncertainty about this led him to state:

> "For while comets move in very eccentric orbs in all manner of positions, blind fate could never make all the planets move one and the same way in orbs concentric, some inconsiderable irregularities excepted which may have arisen from the mutual actions of comets and planets on one another, and which will be apt to increase, till this system wants a reformation." [4]

This reformation or correction of the orbits had to come from somewhere. This led him to integrate his alchemical vital principle and Biblical wisdom with his mathematical knowledge as presented in his

"General Scholium," in *Mathematical Principles of Natural Philosophy* (1687).

> "This most beautiful system of the sun, planets, and comets, could only proceed from the counsel and dominion of an intelligent Being. […] This Being governs all things, not as the soul of the world, but as Lord over all; and on account of his dominion he is wont to be called "Lord God" [Pantokrator], or "Universal Ruler". […] The Supreme God is a Being eternal, infinite, (and) absolutely perfect." [5]

A quote attributed to Albert Einstein, for whom the mystery of nature was not an alien idea, states:

> "The human mind is not capable of grasping the Universe. We are like a little child entering a huge library. The walls are covered to the ceilings with books in many different tongues. The child knows that someone must have written these books. It does not know who or how. It does not understand the languages in which they are written. But the child notes a definite plan in the arrangement of the books – a mysterious order which it does not comprehend, but only dimly suspects." [6]

Werner Heisenberg gave us the famous Uncertainty Principle. It does not refer to some limitation of our knowing/measuring capacity but to an intrinsic ambiguity in reality that cannot be overcome. This point is also made by J.B.S. Haldane:

> "Now my own suspicion is that the Universe is not only queerer than we suppose, but queerer than we *can* suppose." [7]

Modern science, as we know it today, had its beginnings in the Christian West because of a faith that Reason or rational principles could be found in God's creation. Reason is a personal feature, found in Man. A world that is created by a rational being must also possess this personal feature, which we call God. It is possible that an atheistic culture would have never conceived Reason in the world and therefore failed to develop science. It is the task of this article to understand how and why modern science today has turned away from and failed to comprehend this Reason in the world that is similar to the *nous* that Anaxagoras conceived as ruling the world.

Explanation and Correspondence

> "[I]n the Copenhagen interpretation of quantum theory we can indeed proceed without mentioning ourselves as individuals, but we cannot disregard the fact that natural science is formed by men. Natural science does not simply describe and explain nature; it is part of the interplay between nature and ourselves; it describes nature as exposed to

our method of questioning. This was a possibility of which Descartes could not have thought, but it makes a sharp separation between the world and the I impossible.

If one follows the great difficulty which even eminent scientists like Einstein had in understanding and accepting the Copenhagen interpretation... one can trace the roots... to the Cartesian partition....it will take a long time for it [this partition] to be replaced by a really different attitude toward the problem of reality." [8]

Eclipses of the Sun were once predicted using the geocentric epicycles of Ptolemy. They are now described in terms of the heliocentric orbits of Copernicus. Some ancients knew that they could chase away the Moon dog from eating the Sun god whenever they would beat their gongs. Every time they did it, it worked, so they concluded it correlated with the truth.

Each of these examples has something correct or confirming about them, even though they imagine different realities corresponding to them. As it was mentioned earlier, Newton's conception of the solar system considered God necessary to guide the alchemical vitality that was intrinsic to the order and movement in the universe. The mathematical bones of Newton's *Principia Mathematica* were taken by modern physics and presented as a mechanical model of the universe without the Pantokrator. Of course, Newton, himself, wrote his mathematical section as a whole, surreptitiously including his remarks about the Pantokrator only in an appendix or *scholium*. However, the fact remains that observations of the solar system's movements were used to validate both Newton's and modern non-deistic theories although they referred to very different imagined realities.

A map corresponds to an actual terrain and can help one navigate one's way through the real terrain, depending on its accuracy. Yet, the map may never be considered a substitute for the actual terrain since a two-dimensional visual map can never represent the sensed actuality that is experienced in real terrain. A reflection of reality in a mirror may accurately depict the objects being reflected, but one who makes a journey *through the looking-glass* will not discover the real world but a wonderland of exaggerated imaginations, like Lewis Carroll's Alice did.

Atomic Theory and Quantum Theory provide imagined wonderlands that possess some observations or correspondence with true reality. To some degree, each is logical, self-consistent and complete, although Gödel would object to either being at the same time consistent and complete. [9]

If we carefully consider what science is doing here, we discover that anthropocentric or egocentric conceptions of reality – reality as it is *for us* or *for me* – are being erected in place of true reality as it is *by itself and for itself* as described by Hegel. In other words, a subjective conception/theory that is *for us* is being erected as reality *in and for itself*, although it is opposed to objective reality as it is *in and for itself*. It seeks and may have some correspondence with true reality, and if the subjective conception corresponds with the objective reality, the truth is considered to have been reached. This is called the correspondence theory of truth. However, there are problems with this as we noted above, in that different theories may have some correspondence with objective observations and yet still refer to different imagined realities.

The real problem arises when these different Idols of the Mind (man-made images/ideas/conceptions that are for us in our subjectivity) are presumed to be outwardly objective and venerated as the True Reality (reality *as it is* or *by itself and for itself*). Explanations consist of descriptions in terms of the chosen theories assumed as real, even though they are abstractions from the true reality. Since they are abstractions, they never comprehend the concrete reality that they merely represent.

> "Since the creation of the world God's invisible qualities – his eternal power and divine nature – have been clearly seen, being understood from what has been made, so

that men are without excuse." (Romans 1:20, New International Version)

Measurable properties are observations of the true reality that we perceive and incorporate into the realm of our subjectively constructed theory/ideas/idols, which we think allow us to explain the original objects of true reality. Of course, scientific measurement only involves the quantitative superficial outer husk of things, thus it cannot give us a genuine explanation of the essence that courses, for instance, through a blade of grass producing and making it what it is. Such idols cannot comprehend the invisible qualities of the living teleological process at work in the formation of inexplicably complex molecular structures that their models are unable to reconcile. Of course, lacking any semblance of life or consciousness, such mechanical/dynamical models are strictly impersonal and therefore incompatible with any sort of Reason in the world, or personality.

The quote from Romans above admonishes that there is no excuse for such naiveté. It is a failure of scientists to understand what they are doing, and how to properly approach True Reality. Fred Hoyle has remarked:

> "There is a coherent plan to the universe, though I don't know what it's a plan for." [10]

Sir Francis Bacon, one of the earliest fathers of science, warned about creating idols and what should be avoided:

Idols of the Tribe (innate to finite Man) – Deceptions of the subjective mind and imperfect senses are intrinsic to us. Mere imaginations gain the dignity of reality and are mixed with facts so that they become inseparable. Idols are molded from these compounds.

Idols of the Cave (the well of the individual mind) – An individual who is dedicated to some particular branch of learning interprets everything according to the colors of his own narrow field and experience. If the only tool you have is a hammer you treat everything as a nail.

Idols of the Marketplace (semantics or words) – Words make private thoughts public. But when bombastic words are substituted for thoughts one believes he can convince his opponents by out-talking them. This arises from pure vanity that drags the dignity of philosophy and science into the mud.

Idols of the Theater (sophistry) – Putting on a show, arguing in terms of popular familiarities that are false. [11] What is obvious or familiar has not been carefully thought through, thus "what is familiar is not on that account necessarily known." [12]

God, the Universal I, is also Self-consciousness and not merely Consciousness

> ". . . thought is only true in proportion as it sinks itself in the facts; and in point of form it is no private act of the subject." Hegel [13]

Descartes' *cogito ergo sum* ("I think, therefore I am"), has come to signify the slogan for modernity. "I think" puts the all-important power of thinking in the "I". Not only that, it also establishes the absolute being of the I. However, if we try to determine how it is that we perform this amazing activity of thinking, we draw a blank. If I don't know how to do it, then the certainty that I do it should at least seem dubitable.

Hegel looks at the problem more realistically. Withdrawing from reality to be with one's subjective thought alone, is a detachment from the substantial content of True Reality and becomes a conceit or superiority to it. This type of freedom from the content must be given up, and instead of arbitrarily directing the content of one's thought, one's freedom should be sunk into and pervade the content of reality, letting thought be directed and controlled by Reality's own proper nature of which our essential selves are but a part and manifestation. We do not lose anything thereby except our false sense of self (false-ego), and rather gain our true identity and real/concrete freedom.

Free agency does not reside in us as separated from reality, but within true reality itself in which we participate. Reason is in the world, universal rationality that is also intrinsically particular to us individually and to all. By abstractly considering the self as separated from reality and superior to it, we depart from that universal Reason which is the basis for all rational Men to come in accord.

> "Come now, and let us reason together." (Isaiah 1.18, King James Version)

Religion teaches all to surrender to God, the Supreme Reality, from Whom reason and wisdom originate. Wisdom is not a property of the universe, although we find life and intelligence in Nature. It is a quality of a Person, the Divine Personality of Godhead. Just as consciousness and intelligence pervade our body's activities, so God's Personal energy pervades and forms the whole Reality with life and intelligence. Not only does religion teach surrender of our false-egos to the true Reality of God, but also teaches us how to learn the truth by attending to the revelation it, rather than encouraging the tendency to impose and project one's self imaginings onto reality, taking oneself as a separate subjective agent against passive objectivity that lacks its own agency and ability to reveal itself. Reality as possessing personal agency can reveal Godself to us if we adopt the attentive patience that allows the veils of self-centered egotism to dissolve by the practice of meditation, surrender, and service.

Reality as *by itself and for itself*, means that the Absolute is its own origin [cause of itself, *causa sui*] and has its own purposes for itself. Thus Reality cannot be impersonal. To judge good and bad by our self-centered perspectives will not bring us closer to Truth but entangle us further in the misconception of separate interest. This is something each individual has to understand for themselves. To try to force another to this conclusion is itself something that can only arise from misconception and lead one further into delusion.

Experience is the uniting mediator between subject and object, which are the basic aspects of consciousness. Hegel calls the following of the experience of consciousness phenomenology. This leads to the development of true science or science of the True. This is not easy for the abstract analytical modern scientist to comprehend, yet Hegel's *Phenomenology of Spirit* is there for guiding the study of this method.

To study the Whole as a whole and not merely in its parts requires an immersion into that which forms/informs us. The part cannot be understood apart from its unity with the whole. Here we have to deal with the logic of the unity in difference of part and whole. It is not an impersonal abstract unity or oneness that is intended. God is not one with Nature as in Pantheism. God is transcendental to Nature as much as immanent within it.

The soul is transcendental to the body yet immanent in it, as we can observe when the body dies, the soul or life leaves the body behind. The soul is not the body yet at the same time it expresses itself in and through the body. The same soul remains transcendental to the body through all its changes from childhood to youth to old age. In the same sense, God is transcendental to the ever-changing restless universe while remaining the integral universal Mind or Spirit of which the living entities are particular determinate beings. [14]

Some people only acknowledge universal consciousness, and thus end in an impersonal unity as the highest truth. [15] [16] However, this does not properly comprehend the individual (I) and universal (We). This is because, for the individual finite empirical consciousness and the universal consciousness, there exists a true unity and possibility as self-consciousness. The necessity of self-consciousness in every consciousness is explained by Kant in his *Critique of Pure Reason*.

In order for consciousness to be conscious of an object, it must be capable of self-consciousness. Calling oneself I (the ego) refers to this self-consciousness. It is this concept of self-consciousness that Kant called the transcendental unity of apperception that provides the possibility for the consciousness of an object or the unity of the two. Possibility (concept) and actuality (content) are always correlated in this way, so one cannot talk about possibility abstractly (as if it meant anything were possible) without reference to its actuality. Aristotle is famous for stating that the actual is needed for the conception of the possible.

The totality of finite I's, or We, also has its universal essence, i.e., since every I calls itself I there must be a universal I. Thus God, the universal I, is also self-consciousness and not merely consciousness.

Free Will and the Fall

> "PROP. LXVIII. If men were born free, they would, so long as they remained free, form no conception of good and evil.
>
> Proof.—I call free him who is led solely by reason; he, therefore, who is born free, and who remains free, has only adequate ideas; therefore (IV. lxiv. Corollary) he has no conception of evil, or consequently (good and evil being correlative) of good. Q.E.D."
> - Spinoza [17]

By remaining united with the universal Reason of Divine Reality/God the duality of good and evil does not arise. Only when Man withdraws from universal Reason into the particular individuality of subjective understanding (eating from the tree of knowledge of good and evil) is he thrown from the Garden of Eden of unity with God by disobeying His commands (departing from universal Reason).

A truly free individual knows that everything is alright because of one's unity with the divine Reason that rules the world. This freedom from the duality of good and evil and of life and the fear of death is possible for one who's particular will is dovetailed with the universal.

One is free to reject universal Reason and live according to one's own separate egotistic reason and will, but this is the cause of irrationality and evil in the world. The labor that Man and Woman are punished to endure is actually a blessing. For by such work or labor in the world, as much as the labor to properly learn about true Reality, they come to release themselves from self-centered subjectivity and unite with the universal objective Reality that they abandoned due to choosing to cultivate false self-centered understanding withdrawn from universal Divine Reason/God.

I and Mine

> "The truth is the whole. The whole, however, is merely the essential nature reaching its completeness through the process of its own development" (Hegel, 1807, § 20).

The truth is the whole. The part is an abstraction or untruth if it is taken in its isolated identity. It is not what it analytically appears to be. It has its identity only in its integral relation to the whole. To grasp the part from the perspective of the whole a new method is needed distinct from understanding the whole as constituted/constructed from its parts.

The whole has its finite parts. It is not that the part (finite ego) ever vanishes as some Buddhists and nihilists think, or merges into the whole and becomes the whole (abstract monism). The whole always contains its parts, but they are not separated in isolated identity - that is their false identity. Rather, they are understood in relational integrity with the whole. This relation in its highest sense is called love, where there is a lover and beloved on both sides.

Since the part has its true identity as a part of the whole there is a natural affinity of the part for the whole. In its highest or most perfect or satisfying form, this affinity is called love. This can be experienced most fully in human form; although, every partial being, as part of the whole, feels this affinity.

The false I is the sense of identity as a part of the whole separated, unrelated to, or withdrawn from the whole. There is no such separate identity, or false ego, in reality. This false ego is merely brought about by an abstract conception of I and mine. The feeling of having a mind of its own or owning anything arises from this sense of mine. The I is even more troublesome because it thinks "I am" and thus considering itself the whole of being, thinks "I am God." It thinks in terms of its separated self as "I am a human, American, white, etc." But no man is an island unto himself, all are a piece of the whole.

> No man is an island,
> Entire of itself,
> Every man is a piece of the continent,
> A part of the main. [18]

It is false because the true identity of the part is in dialectical unity with the whole - in religious terms as a servant of God. By forgetting/ignoring this unity one's true self or identity is lost. The problem then arises that one thinks he loses his abstract freedom by submission to God, when in fact he gains it concretely for the first time. That's why it is called (*mukti/moksha*) liberation/freedom or salvation in religion.

Freedom requires necessity for it to exist in any concrete sense. If one travels a road without observing rules or laws, chaos, caprice, and destruction would result - not freedom to go where one wants. Such freedom without necessity would be abstract and not real.

The concept of matter or materialism arises when the environment experienced as separate from the divine spirit or God is imagined as having its own separate real being. This creates the imaginary field of separated energy/existence. In the exterior material world, everything is considered external to everything else and separate from everything; everything possesses abstract identity or a false identity only. In other words, each entity is (assumed to be) what it is not, and is not what it is (assumed to be). Aristotle gave the example that a hand detached from the body is not a hand, since it does not serve the function of a hand, which is what we mean by hand.

The whole, from which the false self withdraws, now appears as a transcendental truth for such a person. The soul is one's real identity within the whole. That identity remains undeveloped or dormant because one is developing a purely separated material life, and neglecting their spiritual life. The material body is thus an illusion of a separate being that has its real spiritual existence as part and parcel of the whole or God. When comprehended in true unity with the whole, one's spiritual nature is revealed, although the material misconception initially covers it.

Because the parts of a machine are brought together by an agent or agency outside and apart from the parts themselves, the parts do not of their own accord form themselves into a machine. The parts or members of an organism, however, have their agency within them as the soul which has its identity only in relation to the divine whole.

The analytical understanding cannot deal with integral wholes and therefore cannot understand the soul or God. The material body is illusory in the sense that it cannot be understood in its true identity without knowing its relationship to the whole. Analyzing its composition in terms of separated molecules or neurons is also illusory. Understanding how to go from an untrue, or partially true part, to its truth within the whole, requires a method developed by Hegel called conceptual thinking.

Reproduction is not merely concerned with individuals, it involves the genus or genus process. The universal genus that determines the species is not changed when particulars under the universal change, die, reproduce according to their own kind or species. Without connecting the individual organisms with their species, we may wrongly think they may evolve into any species by chance. But this is not how individuals and species are related to each other. The species determines how the individuals will appear and control their reproduction, growth, and so on. They are not independent to become whatever they want by choice or by chance. The relation between universal and particular is ignored by evolutionists, therefore they cannot develop a proper theory on a purely abstract material basis.

Everyone calls themselves "I" so "I" is universal, although each person means only oneself. In the same way, each organism is identified with its species, although when we speak we refer to a particular organism – for example, a dog and not canine in general. Thus the universal and particular cannot be separated from each other. Similarly, individuals who identify with only personal reason, independent of universal Reason, are misconceiving true reason. Although universal and particular are related, they are not the same, but a unity in difference. The universal by itself is as abstract as the particular by itself. At the same time, their relation can only be comprehended within a higher unity, the higher Self or Supreme Individual.

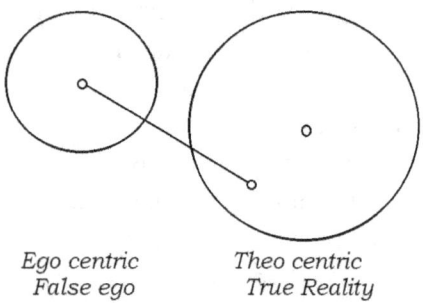

Ego centric
False ego

Theo centric
True Reality

If we were to picture the situation, we would have a circle with God in the center, a theocentric conception, and the finite individual self somewhere inside the circle related to the center. If the finite individual is withdrawn from the circle and placed outside of it with a circle drawn around itself, the egocentric conception, this would represent the materially or mentally conceived world made of illusion or imagination. A line connecting the self in the theocentric and egocentric worlds exists. To re-establish one's true identity one has to give up the egocentric mentality, and its world, and humbly attempt to re-enter the true Reality of Rational theocentric conception and actuality.

Reality Has Its Own Purpose In and For Itself

In an article by Deepak Chopra, M.D., it was mentioned that

"-- The reality we accept is a human construct."

"-- We should see ourselves as conscious creators who imbue reality with our own purposes." [19]

While many people think this way, there are also those who believe that reality is fixed and remains unaffected by our perception of it. To some extent, both are right. We do have the freedom to interpret what we experience, and the mind does play a role in determining what the senses observe. At the same time, reality doesn't just disappear when we do not perceive it. Our house is still there when we go to work for several hours. We assume its permanence but it could burn down when we are not there. So it is not all a human construct, and as for purposes, they also are not solely created by us.

A more satisfactory conception would be one that includes and harmonizes both the idealism of a mind or consciousness-based creation of reality and the realism of the inherent purposefulness of an already existing reality of which we are part and parcel.

The reality people experience is a human construct insofar as it is limited to the sensuous perception of the phenomenal world of appearance, as well as the circumscribed judgments of finite understanding. However, this does not reach to the noumenal Reality in and for itself, beyond or behind its apparent or phenomenal surface. In India, the *mayavad* philosophy of *Brahman Satyam, Jagat Mithyā*, claims that Reality is purely a product of human misconception and only Brahman as mere impersonal consciousness (an oxymoron since consciousness is the essence of personality) is the absolute reality or truth. This philosophy, also called *kevala-advaita*, however, does not provide an alternative to material reductionism but merely an alternative reductionism. Instead of reducing everything to matter, it proposes to reduce everything to impersonal consciousness. In fact, what is needed is an alternative way of thinking that is not based solely on the judgments of a finite ego that is found, for instance, in the abstract thinking of Kantian philosophy in the West as well as in the *kevala-advaita* interpretations of Śrī Śaṅkarāchārya in the East.

If we consider that Reality already and always exists by and for itself then Reality or its purpose does not have to be created by humans. Rather humans are one of the many products or creations of a Reality that exists for itself, i.e. as self-conscious being for itself or having its own purposes. Because we are part and parcel of a self-conscious Reality we are also conscious. Because Reality is also by itself (or in itself), it is substance. Thus, Reality is or exists as, a self-conscious substance, and therefore finite instances of it are conscious substances or thinking beings. Yet these instances are not all at the same level of consciousness, but fall within a spectrum of consciousness manifested as different forms of life. This is because Self-conscious Reality is not abstractly or one-sidedly monistic or pure oneness, but is itself differentiated within itself having many qualities
or what is called Personality. Personality is self-consciousness that is a oneness, or Individual, that also contains varieties or differences within it.

Most people born in India within the Hindu tradition have heard of *sanatana dharma*. It means there is already an eternal purpose (order) in Reality that is neither created nor destroyed at any time. According to the Bhagavad-gītā 2.16, whatever is created is temporal - only a fleeting reality, like a dream. Thus the purposes humans create are just dreams, while the eternal purpose or *sanatan dharma* is the universal order or purpose valid for all creation and creatures regardless of their individual purposes. The Bhagavad-gītā explains that living solely according to self-created purposes, and not in accord with *sanatan dharma*, is called *Māyā* or illusion.

It may be more difficult for atheists to comprehend that Reality has its own order, purpose or being-for-itself, because Reality for them may be understood merely as something impersonal and indifferent. Yet, if the Ultimate or Absolute Reality is a sentient substance (Thinking Being) it must have will or purpose. That is not the same as the impersonal reified laws that scientific discovery seeks as the universal or intelligible laws of Nature. However, what we may call Divine or Infinite Reason that belongs to Reality is not the same as the finite reason that the human products/instances of Reality may create for themselves. The latter are living like bubbles (with themselves as the center) on the ocean of Reality or Truth. This is the nature of *Māyā*. However, we are not apart from Reality but an implicit part of it. A very different attitude toward Reality is required to understand that difference.

The consequence of the conception that we create our own reality is already producing its results all over the world. When the bubbles of individuals or groups collide with each other the result is jealousy, war, hypocrisy, and strife. This is what philosopher

Thomas Hobbes (1588-1678) called "the war of all against all" (*L. bellum omnium contra omnes*). [20] This egoist conception is based on the limited idea of the being of Reality for one's own finite consciousness, that misses the universal being-for-self of Reality for its own self, which is infinite or the same for everyone and everything. Thus, even if we accept the principle that we create our own reality (according to whatever makes us happy), the idea of individually centered realities leads to destructive consequences in the real world. The Bhakti Vedānta alternative is for each to create or understand an unselfishly directed Theo-centric Reality with its own purpose that can harmonize the freedom of each individual to produce a harmonizing Reality in which all may live peacefully. In other words, creating your own reality doesn't have to be selfishly oriented which leads to an ultimately false or illusory happiness, but Theo-centric which leads to humility, love, and a peaceful harmonious life.

In order to regain their connection with Reality, one must burst their own bubble of self-centered reality and surrender to their eternal constitutional purpose or *sanatan dharma*, which is not to be identified with any political, social, or other partisan religious groups. Rather, it is that purpose which belongs to all such groups, individuals, and even inanimate instances of Reality within the cosmic and trans-cosmic order.

You may rightly inquire what the universal purpose is that the individual must dovetail with the universal Will. Science, when properly conceived, is an attempt to discover the order or laws that implicitly govern Nature. However, instead of deriving those laws from the observations of Nature, modern science has retreated from Nature into their own theories, ideologies and models of Nature. Thus, rather than concluding from the natural observation that life comes only from life, they create an ideological scenario in which they have now forced themselves to think that life comes from matter. This completely opposes what is observed in Nature.

The rational study of Nature is not the only way to understand the universal Will. Those wise and saintly souls who have plumbed the depths of Reality also have something to say about these things. We can learn from them if we have the intelligence to understand their contribution to human knowledge, which concerns the most profound spiritual nature of Man beyond the immediate surface of appearing Nature. Those great saints and sages, spiritual scientists, have recorded their discoveries in revered books called scriptures or revealed knowledge. Their understanding of true knowledge or knowledge of the truth is that it is always self-revealing, being the inherent constitutional nature of one's own true self and the cosmos.

As Krishna, the Name of the self-revealing Reality, explains at the end of the Bhagavad-gītā 18.66, all individually motivated *dharma* or purpose must be renounced (*sarva-dharmān parityajya*) and one must surrender (*śaraṇaṁ*) to Him, the self-conscious Reality that the various religions call God. This surrender entails submitting one's individual will to serve the interest or purpose of the universal Will, under Whose direction the entire cosmos moves and derives its existence and purpose. Opposed as this may seem to the modern idea of scientific thinking, this can be demonstrated to be a completely scientific and rational conception from all points of view that we attempt to present. Although it is a revolutionary way of thinking, surrendering to Universal will provides a rational alternative that can be justified only after careful study and application of the Bhakti Vedānta philosophy.

References

[1] Dobbs, B. (1991) *The Janus Faces of Genius*. Cambridge University Press, pg 7.

[2] Brewster, D. (2010). *Memoirs of the Life, Writings, and Discoveries of Sir Isaac newton*. Cambridge University Press.

[3] Chenciner, A. (2012). Poincare and the Three-Body Problem. Retrieved from http://www.bourbaphy.fr/chenciner.pdf

[4] Halsall, P. (Ed.). (1997, August). Modern History Sourcebook: Isaac Newton: Optics. Retrieved from https://sourcebooks.fordham.edu/mod/newton-optics.asp

[5] Newton, I. (1729). Andrew Motte's translation of the General Scholium to Isaac Newton's Principia (1729). Retrieved from https://newtonprojectca.files.wordpress.com/2013/06/newton-general-scholium-1729-english-text-by-motte-letter-size.pdf

[6] Gaither, C. C., & Cavazos-Gaither, A. E. (2012). *Gaithers Dictionary of Scientific Quotations: A Collection of Approximately 27,000 Quotations Pertaining to Archaeology, Architecture, Astronomy, Biology, Botany, Chemistry, Cosmology, Darwinism, Eng.* Springer, pg 1419.

[7] Haldane, J. B. S. (1927). *Possible Worlds and other Papers*, pg 286.

[8] Heisenberg, W. (1962). *Physics & Philosophy: the Revolution in Modern Science*, pg 81.

[9] Raatikainen, P. (2015, January 20). Gödel's Incompleteness Theorems. Retrieved from https://plato.stanford.edu/entries/goedel-incompleteness/

[10] Knowles, E., & Partington, A. (1999). *The Oxford Dictionary of Quotations*. Oxford University Press.

[11] Hall, M. P. (n.d.). The Four Idols of Francis Bacon. Retrieved from http://www.sirbacon.org/links/4idols.htm

[12] Hegel, G.W.F. (1807). Phenomenology of Mind/Spirit, § 31. Retrieved from http://www.gwfhegel.org/PhenText/compare.html

[13] Hegel, G. W. F. (1975). Science of Logic, §23. Retrieved from https://www.marxists.org/reference/archive/hegel/works/sl/

[14] Sridhara, B. R. (n.d.). Srimad Bhagavad Gita, Ch. 13 Distinction between Matter and Spirit. Retrieved from http://gaudiyadarshan.net/index.php?b=91&c=18&m=0&blc=0&vlc=0&l=1

[15] Kafatos, M. C., & Yang, K.-H. (2016, December). The quantum universe: philosophical foundations and oriental medicine. Retrieved from https://www.ncbi.nlm.nih.gov/pmc/articles/PMC5390421/

[16] Senese, M. (2017, May 3). An Interview with Deepak Chopra: You Are the Universe. Retrieved

from https://www.kosmosjournal.org/article/an-interview-with-deepak-chopra-you-are-the-universe/

[17] Spinoza, B. de, & Elwes, R. H. M. (2018). *The Ethics*. Mineola, NY: Dover Publications, Inc.

[18] Donne, J. (1959). Devotions Upon Emergent Occasions, XVII. Meditation. Retrieved from https://www.gutenberg.org/files/23772/23772-h/23772-h.htm

[19] Chopra, D. (2018, October 9). Can We Evolve Beyond Evolution? We Have To. Retrieved from https://www.sfgate.com/opinion/chopra/article/Can-We-Evolve-Beyond-Evolution-We-Have-To-10818947.php

[20] Thomas Hobbes. (2019, June 28). Retrieved from https://www.biography.com/scholar/thomas-hobbes

The Unreasonable Effectiveness of Mathematics in the Natural Sciences

The Identity of Indiscernibles

Leibniz' principle of the identity of indiscernibles states that there cannot be separate objects that have their properties in common. [1] Thus 1 and any other 1 in mathematics being exactly identical cannot refer to physical objects. No two objects in physical nature are exactly identical, thus mathematics has a different logic than Nature. As Wigner implied in his "Unreasonable Effectiveness of Mathematics in the Natural Sciences," it would be unreasonable to think otherwise. [2]

Counting one apple, two apples, and so on, is not the same process as counting numbers, $1 + 1$, etc. First, we note that all 1's are exactly identical in arithmetic, but no two apples are identical in physical nature. When we count apples as if they were identical entities we are not referring to actual apples, but a notion, or sortal, we call "apple". This distinction is generally ignored. The point is that counting numbers in mathematics and counting physical objects deal with different entities.

Galileo's Error of Primary and Secondary Properties

Galileo Galilei (1564-1642) proposed that mathematics was the language of the book of Nature, innovating the use of mathematics in what has become modern science. What is usually not explained is that Galileo had a particular view of reality that made this seem possible. His perspective was that physical objects had what were called primary and secondary properties. Primary properties were quantitative, such as size, weight, and so on. Secondary properties were qualitative phenomena, such as taste, color, and so on. These secondary qualities belonged to the sensory organs perceiving physical objects but not to the objects themselves, which possessed only primary properties. Thus color, for example, was determined by the sense organs of a human being, or other creature, and not by the object being perceived.

Of course, that is not the way we understand physical objects today. Primary quantities and secondary qualities of physical objects are thought to belong to the objects themselves, and not divided between the observers and observed. Some have interpreted this to mean that Galileo thought that phenomenal qualities belonged to consciousness and quantities to physical objects. [3] However, it has been claimed that he meant the sense organs rather than consciousness, although it may be argued that the sense organs do not sense anything without consciousness. The intact organs of a dead body do not sense anything when life or consciousness is missing.

The problem with dividing sense perceptions from objects is, of course, that sense perceptions have to come from somewhere. One doesn't just experience some quality without its relation to the object whose quality it is. A more integral metaphysics is needed to heal the division that was promoted by Galileo and others. The mathematization of physical nature was based on a metaphysics that is divisive and incomplete.

Philosophical Development of Ideas that Science Ignores

Descartes (1596–1650), independent thinker that he was, similarly held the dualistic view of qualitative and quantitative measures, and forged the idea of the numerical nature of space and time. For him, mind (and metaphysics) was to be kept separate from the extended (spatial) physical world of matter (physics).

Locke (1632-1704) argued that objects characterized by size, shape, and so on could not be distinguished from a region of space of the same size, shape, and so on. Berkeley (1685-1753) explained that primary qualities are determined by experience as much as an object's qualities. This undermined the whole metaphysics of Galileo and others, yet the dogmatic applicability of mathematics to science maintains a foothold in science to this day, without metaphysical critique.

While Berkeley tried to bring the idea of the world into the subjective mind, Kant (1724-1804) attempted to reinstate the denuded objectivity of things in terms of an abstract thing-in-itself that was unknown and unknowable but logically posited by thought. Schelling (1775-1854) furthered this historic development in philosophy by offering that the subjective and objective worlds were perspectives that were identical in their difference, thus overcoming the strict divide between them.

Hegel (1770-1831) then brought the whole development to an absolute conclusion by explaining that the idea of the world was not in a particular subjective mind alone, but there is also the idea of the objective world within which the particular observer is included. This leads to the necessity of a comprehensive Absolute Idea that has being-in-and-for-itself, in which both the particular subjective and universal objective perspectives are dynamic participants of a higher unity in difference that preserves yet sublates them in a negative self-conscious individual unity.

From Pantokrator to Chaos

Many other problems exist in applying mathematical mechanics to Nature. Newton (1642–1727) had to invoke a Pantokrator to correct for the unavoidable disastrous perturbations that the planets in the solar system would suffer in their journeys. [4] The ancients believed the planets followed ritual performances (*Rta*) but rejected the idea that they were being constrained by force like naughty children. From the history that we now call myth, we know in both the Occident and Orient that the ancients long held that the planets were free spirits, rational personalities, as subject to laws - but not impersonal mathematical formulas.

The problems created by mathematical mechanistic theories in cosmology have led to a crisis in the deterministic model that has ruled scientific thinking in the modern period. It has given rise to the theory of Chaos that is now the reigning concept of the Universe, which cannot simply be swept under the carpet by those who still support the doctrine of mechanistic science. In a 1986 speech, James Lighthill said,

> "We are all deeply conscious today that the enthusiasm of our forebears for the marvelous achievements of Newtonian mechanics led them to make generalizations in this area of predictability which, indeed, we may have generally tended to believe

before 1960, but which we now recognize were false. We collectively wish to apologize for having misled the general educated public by spreading ideas about the determinism of systems satisfying Newton's laws of motion that, after 1960, were to be proved incorrect". [5]

Many scholars agree that the argument for a clockwork universe as a strict consequence of Newtonian dynamics is no longer logically valid, although it is still held to be true by many in the general public and among many scientists. Due to complexity and errors accumulating exponentially over time, we cannot be certain of determinism during short intervals, in principle, or even for classical systems. Nature seems to draw a curtain on predictions of mechanical motion in a clockwork universe that is beyond our ability to unveil. [6]

Logic of Mathematics Compared to the Logic of Nature

Einstein once remarked,

"Everyone who is seriously involved in the pursuit of science becomes convinced that a spirit is manifest in the laws of the Universe - a spirit vastly superior to that of man, and one in the face of which we with our modest powers must feel humble." [7]

There should be a question mark between mathematics and Nature because they deal with two different logics. During a speech given in 1921 at Berlin's Prussian Academy of Science, Einstein addresses this relationship:

> "One reason why mathematics enjoys special esteem, above all other sciences, is that its laws are absolutely certain and indisputable, while those of all other sciences are to some extent debatable and in constant danger of being overthrown by newly discovered facts."

He then continues:

> ". . . it is mathematics which affords the exact natural sciences a certain measure of security, to which without mathematics they could not attain.
>
> At this point, an enigma presents itself which in all ages has agitated inquiring minds. How can it be that mathematics, a product of human thought which is independent of experience, is so admirably appropriate to the objects of reality? Is human reason, then, without experience,

merely by taking thought, able to fathom the properties of real things?

In my opinion, the answer to this question is, briefly, this: As far as the laws of mathematics refer to reality, they are not certain; and as far as they are certain, they do not refer to reality." [8]

In other words, mathematics works within its own framework of certainty based upon a set of axioms of mathematics thus yielding specifically determinable results. There is, however, no reason to expect that the reality of Nature obeys such axioms or that observations and experiences of Nature should in any way correspond to the realm of mathematical logic. Thus the employment of mathematics in science is more or less just to give the scientists comfort and security rather than any real or ontological connection between mathematics and science. Furthermore, Gödel did much to dispel even the certainty that mathematics falsely claimed for itself. [9]

So far as the touted accuracy of quantum theory is concerned, it must be understood that even false scientific theories can give amazingly accurate results. In the past, the Chinese beat a gong to make the Moon Dog stop eating the Sun god during a solar eclipse - the theory works flawlessly 100% of the time. That quantum mechanics works so accurately for a single electron or even two particles, ignores the fact that, as a mathematical construct, there is no such thing as a single electron or two in the universe. The accuracy of the whole theory breaks down when we consider any real life-situations without idealized approximations.

It may be uncomfortable for scientists to think about these difficult but very real problems that afflict the very foundations of modern science, but it does not alleviate or eliminate such problems by ignoring them. Positively, they provide an incentive to rethink our conceptions of science - making not only the known but also our knowledge of the known an object of study. This will raise science to self-conscious science, include the scientist in science, and change the method of science in the process.

References

[1] Forrest, P. (2010, August 15). *The Identity of Indiscernibles.* Retrieved from https://plato.stanford.edu/entries/identity-indiscernible/

[2] Wigner, E. (1960, February). *The Unreasonable Effectiveness of Mathematics in the Natural Sciences*. Retrieved from https://www.dartmouth.edu/~matc/MathDrama/reading/Wigner.html

[3] Goff, P. (2019). *Galileos Error: Foundations for a New Science of Consciousness*. New York: Pantheon Books.

[4] Newton, I. (1729). Andrew Motte's translation of the General Scholium to Isaac Newton's *Principia* (1729). Retrieved from https://newtonprojectca.files.wordpress.com/2013/06/newton-general-scholium-1729-english-text-by-motte-letter-size.pdf

[5] Debnath, L. (2008). *Sir James Lighthill and Modern Fluid Mechanics*. London: Imperial College Press, pg 31.

[6] Snobelen, S. D. (2012, February 8). *The Myth of the Clockwork Universe*. Retrieved from https://isaacnewtonstheology.files.wordpress.com/2013/06/the-myth-of-the-clockwork-universe.pdf

[7] Einstein, A., Hoffmann, B., & Dukas, H. (2013). *Albert Einstein, The Human Side*. Princeton University Press.

[8] Einstein, A. (1921, January 27). *Geometry and Experience*. Retrieved from

http://mathshistory.st-andrews.ac.uk/Extras/Einstein_geometry.html

[9] Raatikainen, P. (2015, January 20). *Gödel's Incompleteness Theorems*. Retrieved from https://plato.stanford.edu/entries/goedel-incompleteness/

The False Elephant and the False Ego

When we look at the world through a microscope we see only cells and their internal moieties - we never see people, clouds, trees, or elephants. Even the macromolecular vision of such separated entities does not show us the purpose or whole of that for which each exists. Such is the nature of the analytic thinking of finite cognition that we call modern science.

The blind men who could only feel the different limbs of an elephant concluded the legs were tree stumps, the side was a wall, the tusks were spears, the trunk a serpent, the ears a fan, and the tail a rope. Not knowing the unity of the whole they could not know the relation of each part to it. Without that knowledge, they could not have a proper understanding of the parts in isolation from the whole. Even a simple understanding of the function of each part in isolation could not reveal the relation of the functions to each other, what to speak of their relation to the unity of the whole of which they had no clue.

Aristotle said that there are four factors which must be comprehended in order to explain things. These were the material, efficient, formal, and final cause. These four represent what a thing is made of (material), what modifies it (efficient), what plan/design guides its progressive modification (formal), and the goal - the end or aim as the final product which corresponds to its concept (final end or purpose).

In reality, it is the final end that determines and guides all the other factors. When the unity of the whole is sentient or self-conscious, even a systems approach will not be sufficient to explain its existence. Systems are not self-conscious.

Without knowing the purpose for an individual's own self-conscious existence one has only a false conception of self (false ego). One's true identity is established only when it is comprehended in relation to a proper concept of the Complete Whole. This Complete Whole is the all-accommodating, all-pervading, all-unifying, all- attracting, self-conscious, and self-knowing Absolute Truth, Who is known by many Names.

Unity of Science and Religion

The Link Between Science and Religion

Science and religion are the two major subjects that people turn to for deepening their understanding of life. Some people consider that science addresses the "how?" of life, while religion addresses the "why?". Although these vague considerations may not directly assist the deeper rational development of either of these subjects, they hint at the vital role that thought and consciousness have in our relationship with truth. If we understand thought to be only that of which consciousness is aware, then one's own thoughts about what consciousness is must also be only objects of consciousness and we have not yet reached what consciousness itself is. Considering both consciousness and its content (objects, thoughts, intuitions, etc.) in which the content specifically refers to thought or thinking, we have to first examine the uncritical presumption that consciousness is merely different from and opposed to its thought-content.

However, instead of the belief that consciousness is passive and opposed to thinking, which is active, we may consider the idea that thinking is the activity of consciousness. In that case, the opposition between the two is resolved as an integral and dynamic unity. In other words, we have a unity of unity and difference, or the passive and active in one dynamic whole. This is similar to the subject-object unification that comes about by recognizing that an object is what a subject knows it to be.

Importance of Concepts and Their Content

It is very difficult for the modern mind influenced by empirical science to grasp what is being offered here. This is due to the type of reasoning or epistemological method that characterizes empirical science (called positivism), which holds that only the positive sense perceptions contribute to knowledge, with no place for the negative or subjective thoughts and concepts that are essential counterparts of sense perceptions. In other words, the contribution of concepts (rationally developed thoughts) is basically ignored. As a result, the ability to think conceptually has been practically lost due to attrition or atrophy. Concepts in rational thought become hypotheses, theories, paradigms, models, probabilities, and so on, in modern empirical scientific thought. In other words, rational concepts have become replaced by contingent guesses or correlations because the dynamic unity in difference between concepts and their content, or between thinking and things has not been properly comprehended. This is a serious and fundamental deficiency of modern science, having the deeply troubling consequence of crippling humanity's ability to think properly.

We have to try to grasp what the ancient and modern philosophers have already explained about this topic. Generally, we will find that the greatest difficulties for proper understanding lie in the uncritically accepted differences that we presume to be beyond doubt, such as the subject - object duality. What may seem to be completely revolutionary could turn out to be the key to overcoming whatever impasse that may be blocking our proper comprehension.

In addition to the unrecognized dynamic unity of the moments of concept and content, or thinking and being, there are simultaneously differences involved in this relationship. When considered in mere opposition, duality seems to represent no relation, but when each aspect of duality is considered as relatable to the other, we arrive at what German philosopher G.W.F. Hegel called "being-for-self." This is another major category missing in scientific discussions. Hegel explained "being-for-self" as the logical category of the being of consciousness, an inherent subject-object structure. He offered that the general notion of being may be further specified as "being-in-(or by)itself" as well as "being-for-itself." This is useful in analyzing knowledge which involves:

1. The being in itself of the object to be known
2. Knowledge of the object (the being-for-consciousness of the object)
3. The knower (being-for-self)

Knower: Knowledge: Known

The known object has its own being by itself. The being of the object for consciousness is called knowledge of the object. What the object is for consciousness, we call knowledge, and that for which knowledge is, the knower. The self (*atman*) is thus connected with the object through his/her knowledge of it. We may call the study of the 'known' ontology, the study of knowledge epistemology, and the study of the 'knower' theology. All three of these are integrally connected, but modern science only deals with the mind-independent known. By strictly focusing on the object known, modern science does not explicitly account for the role of knowledge (being-for-consciousness) involved in what is known. In other words, it does not make knowledge an object of its investigation. Yet because the object is what the subject knows it to be, there is an inseparable connection between subject and object that is lost when the focus is on knowing the object alone. Thus when it comes to examining the relation between the knower (subject, consciousness) and the known (object) we must necessarily go beyond the mere scientific study of the object involving the category of being in itself (by itself) and take up the study of a different category of being, namely being-for-consciousness or being-for-self. A scientific study (meaning rational and systematic) can be undertaken in this case involving categories of being beyond mere objects, and this has been demonstrated in Hegel's *Science of Logic* [1] and *Encyclopedia of the Philosophical Sciences.* [2]

It is important to emphasize this point. The known is what we have some knowledge about, but science neglects the role of the scientist's knowledge or the being-for-consciousness of the object. The relation of these three aspects (knower, knowledge, known) is something philosophers have always considered essential for complete comprehension of truth, but scientists ignore everything but the known object.

Immanuel Kant, a German philosopher who is considered a central figure of modern philosophy, made especially clear the contribution of the subject, the knower (or mind), as essential to understanding the unity of the object and its properties. It is not that the object can be understood on its own. There's an integral contribution from the subject in determining what the object is. A result of the failure to study his philosophical arguments in this regard has led to much confusion in the development of quantum theory concerning the role of consciousness.

The other consideration of 'being-for' that is neglected by modern science is the purpose, or that for the sake of which an object exists. This is interpreted as the final cause in Aristotle's philosophy. However, Aristotle's original term refers more to explanation [*aition*] than to cause. To explain a thing properly requires knowing that for the sake of which it exists. The famous example he gave is of a hand. Without knowing the relation of the hand to the human body we cannot properly describe it as merely a five limbed object.

Four Aspects or Types of Cause

This neglect of the other categories of being, which are necessary to properly explain reality, begins with a philosopher-scientist named Francis Bacon (considered the father of scientific empiricism). He proposed that science should be concerned with two of the four factors that were given originally by Aristotle for properly explaining things. Aristotle enumerated four aspects of every cause – the material, efficient, formal, and the final cause. For example, the material cause may be a lump of clay, the efficient cause is the potter's hands molding the clay, the formal cause is the design or plan by which the potter directs his actions and the final cause is the adequacy of the finished product to its idea. Bacon considered that science should only be concerned with two things: the material cause (the material we are working with) and the efficient cause (what causes that material to change) - nothing else. He neglected to consider the formal and final causes. Thus, modern science has inherited that underdetermined form of explanation in terms of material and efficient causation.

The best way to understand this is through the analogy of the blind men and the elephant. Once, there were several blind men and an elephant. The men could not see and did not know that there was an elephant before them. One touched the leg and said,

"Oh, this is like a tree stump," another touched the tail and said,

"Oh, this is a rope," a third man touched the trunk and said,

"This is like a snake," and a fourth man touched the ear and said,

"This is like a fan." Because they did not know the whole, their knowledge of the parts was incomplete and consequently improperly understood. Without understanding its purpose or relation with the whole, a leg of an elephant could be misconceived as a tree stump. Without knowledge of that for the sake of which a thing exists, it can take on a completely different significance altogether. So, by studying the different finite aspects of the world without understanding their overall purpose, we cannot properly understand anything. We don't have the full information we need to properly explain things. Thus, being-for, or understanding the purpose of life, has become something many people no longer ask anymore. They think that the purpose of life is simply to enjoy the senses and that everything is finished at the time of death. That's it! This whole way of thinking can trace its expression in modern science.

Descartes: Subject-Object Duality of Consciousness

Modernity refers to the age of reason or enlightenment philosophy, which was first clearly formulated by Rene Descartes (1596-1650). Here, we first find consciousness (*res cogitans*) or subjectivity being distinguished from matter (*res extensa*) or objectivity. This gave rise to the concept of the duality of the subject and object.

But what is cognition? What does knowing mean? How do we raise what is externally objective (being-in-[or by] itself) to our own subjectivity (being-for-self)? What we understand as subjective is an internal image (picture) or representation of what is outwardly present in the world. In philosophy, this subjective representation of phenomena is called perception. A perception is something like a preconception or something that comes immediately, before forming a proper conception or comprehension of an object. Thus we can distinguish three stages present in consciousness: sensuous apprehension, perception, and comprehension.

Each form (stage, level, Gestalt) of consciousness can thus be distinguished from the others by its particular function. The first level of consciousness is the senses or sensuousness, also called sentience. Specifically, that which can feel or experience is sentient. To claim that the Absolute is a sentient being or sentient substance is to recognize its ability to feel and experience. This completely revolutionizes the presupposition of modern science that reality is ultimately impersonal, insentient substance or matter. The materialized conception of the universe finds it impossible to explain how sentience or life came from insentient matter. If life or sentience is primordial, however, the hard problem that materialism faces is immediately resolved. This idea correlates with our natural experience that life always comes from a previous life. To explain how the concept of a material rock is formed, for instance, seems much more reasonable than how a rock could form a concept of itself using a scientist for that purpose. This point is also recognized by Nobel laureate George Wald, as seen in his statement, "A physicist is an atom's way of knowing about atoms." [3] So, the immediate apprehension of an object by the senses may be considered the first level of consciousness found even in lower forms of life, such as microbes, plants, and insects, who may not have the developed bodies of a man or animal but nonetheless do exhibit behavior indicative of sentience or feeling.

Levels of Consciousness

The next level of consciousness is the perception or the pre-conceptual formation of internal representations, images, or pictures. These may also be considered as immediate intuitions of the internal sense, but because representations are reflected mirror images of external objects, they are considered the mediating means between apprehension (the first level of consciousness) and comprehension (the third level of consciousness) or the concept of the representation that is formed by thought. It is within the hierarchy of these three initial levels that we can recognize the beginning of what is called the subjective evolution of consciousness.

Kant took this simplified and naïve schema of the stages of cognition and subjected it to his method of internal critique – thinking about, analyzing, and systematizing the internal processes of the mind. He recognized that there are certain inherent categories of the mind that are applied to the objects that are intuited by the senses. In other words, the objects of the senses are not just reflected in the mind, like an internal mirror, and then we arbitrarily form concepts of them. Rather, the mind has to apply certain judgments to the objects that the senses apprehend to determine them as unities, differences, relations, and so on. The senses understood as mere passive sensors or receivers do not determine, for example, something as simple as object A is 'next to' object B. In other words, none of the passive senses have the ability to determine 'next to.' This is an act of comparison or judgment about what the senses apprehend in a direct, intuitive, simple, immediate, or unmediated way. This means that the mind contributes in a very important way to what the senses apprehend. The naive realists (people in general, including most scientists) have no idea of this involvement of the mind in their perception of the world.

Kant enumerated twelve categories of judgment [4] that were first given by the amazing intellect of Aristotle, about two thousand years before Kant. These are all necessary for explaining our cognition or knowledge of the varieties of phenomenal experience, mathematics, science, and so on. To gain clearer understanding, we may consider the example of robotics, in which a robot is equipped with a photoelectric tube as a sensor – analogous to the eyes that see things. When the sensor detects a photon bouncing off an object, the photoelectric cell generates an electric current. In order for the robot to respond to that current, it has to be passed to a computer that is programmed to send an electrical signal to a servo motor that can move the robot away from or towards the object. In the same way our senses send signals to the brain (through the optic nerve and occipital region) but the mind has to determine how to interpret those signals as to the nature of the object and one's behavior toward it.

Science Has Lost Its Self-consciousness

Modern science has pretty much ignored all of this internal processing in its study of external Nature. Despite its successes, this shortcoming neglects the wealth of knowledge within us. Western philosophy, as well as the ancient philosophy of India, recognized the importance of the internal or subtle processes of cognition in explaining the phenomenal world. Besides the senses, there's the mind, the intellect, and the ego (or unity of apperception, as Kant called it). Memory is also very important. These are part of the subtle reality that is absolutely essential if we want to properly explain our knowledge of things.

From a particular subjective perspective, the totality of experiences in life/consciousness are integrated by a unity called ego. At the empirical level, this unifying agent is considered the false ego. As being-for-self, this false ego fails to account for its integrity with or being-for the universal whole of the reality of which it is part. In other words, the ego is not merely its being-for-itself, it also has being-for that which is other than itself. By dividing itself as ego from the non-ego or world, as self and not-self, it creates a subjective duality that does not exist from the perspective of the whole as an organic unity. Thus it is called the false ego. However, the being-for-self of the particular does exist as a moment of particularity of the universal whole, without being abstractly isolated from it.

There is another way to understand this from the analogy of the elephant and the blind men. The misconception of the elephant's leg as a tree stump arises from the failure to understand the particular object of immediate experience in relation to the whole. Thus the false conclusion that it is a tree stump is due to the failure of accounting for its being-for that which is other than itself – in other words, that for the sake of which it exists, i.e. its final purpose or end. Such limited or finite understanding of objects as isolated on their own must lead to this type of false knowledge.

Isolated and Independent False Ego

Until we understand what we are for, what the purpose of life is then the I, the self, the ego, is misconceived as an isolated self, an individualized form centered upon itself. It is called false ego because it is understood only in its being-for-itself and not in its context as being-for-other, in its relationship to the Organic Whole (God, Spirit, the Absolute, the ultimate truth). In these terms, it becomes real knowledge of self. Our real identity is as part of the Whole. We are part of something greater than ourselves beyond our localized existence, and yet the whole is not merely beyond or opposed to oneself since one is included in the whole. Real education is meant to learn about this.

When a person says "we," one immediately takes oneself and expands it to become a part of something wider than oneself. I and we are not the same (singular and plural first person), yet one still identifies oneself with we. In other words, I include myself when I say "we." I don't think that it is something different from me when I say "we." When Jesus offered the faithful the *Lord's Prayer*, He began with "*Our* Father," not *My* Father. The American Declaration of Independence begins with "*We* the people." Although it often goes unrecognized, it is very important to understand the distinction and relationship between the universal We and the particular I. A numerical and categorical difference is involved yet there is an identity at the same time. In set theory, it would be like the difference between a member of a set and the set itself and their relation. At the same time, from the vantage point of Life in general, "We" may include only human beings - although animals, plants, and even the Earth itself are all essential and inseparable parts of Life.

The Principle of Bhakti

In Vedānta philosophy, the central concept of Bhakti refers to the yoga of the heart - the yoga of devotion (being-for-other). The different processes of sacrifice (*yajna*) or devotion (bhakti) offer the boon of decentering the self. The processes for decentering the self, or dissolving the self-subsistent ego, are fundamental to the Vedic tradition. The perfection of being-for-other is achieved when it takes the form of divine love or Krishna *prema*. This conception requires understanding that Truth as involving consciousness, means that the Truth is personal. Western philosophy uses the Latin *amor dei* to designate this conclusion. Thus it is possible to find the unity of science and religion if we are willing to look more deeply into the logical and rational nature of these superficially apparent disparate subjects.

References

[1] Hegel, G.W.F. "Science of Logic." *Hegel's Science of Logic*, www.marxists.org/reference/archive/hegel/works/hl/hlconten.htm.

[2] Hegel, G.W.F. "Encyclopedia of the Philosophical Sciences." *Contents of Hegel's Encyclopaedia of the Philosophical Sciences in Outline*, www.marxists.org/reference/archive/hegel/works/ol/encycind.htm.

[3] Wald, George. "Life and Mind in the Universe." *George Wald: Life and Mind in the Universe - IV. The Evolution of Consciousness*, www.elijahwald.com/lifeandmind.html.

[4] Kant, Immanuel. *Critique of Pure Reason*. Translated by J. M. D. Meiklejohn, 2nd ed., Everyman's Library, 1984.

History of the Princeton Bhakti Vedanta Institute

The original Bhaktivedanta Institute (BI) was founded by Sripad Bhakti Swarup Damodara Maharaja, under the direction of His Divine Grace Srila A.C. Bhaktivedanta Swami Prabhupada in 1976. The BI's mission was to conduct fundamental research into the nature of consciousness and self, inspired by the ancient Indian wisdom of Bhagavat Vedānta, as taught in the lineage of Sri Krishna Chaitanya Mahaprabhu and His Divine Grace Srila Saraswati Thakura (pictured above). Srila Prabhupada hoped to expose the misconceived speculation and false doctrines that are promoted under the banner of scientific advancement. [1]

The founder and current Serving Director of our Princeton Bhakti Vedanta Institute, Bhakti Madhava Puri, Ph.D., was one of five founding charter members of the BI. In cooperation with the Founding Director, Bhaktisvarup Damodar, Ph.D., B. Madhava Puri became the Regional Director of the Bhaktivedanta Institute's center in Juhu Beach, Mumbai, India, in 1976. During the 1970s, he met with and initiated the profound transformation of

Nobel laureate Professor George Wald by suggesting that he consider how Life comes from Life, as opposed to Life coming from Matter.

Professor Wald (left) is pictured below, walking with Bhakti Svarup Damodar (middle) and Bhakti Madhava Puri (right), at the Bhaktivedanta Institute in Juhu. George Wald's astonishing change in heart and mind can be seen in the following two quotes:

"In 1952 I was giving the Vanuxem Lectures at Princeton University on the origins of life and biochemical evolution. Albert Einstein, whom I had come to know, was walking with me before the first lecture and asked, 'Why do you think the natural amino acids are all left-handed?' ... All I could think of was: the negative electron won in the fight. I said, 'That is exactly what I think of those left-handed amino acids – they won in the fight.'"

"A few years ago it occurred to me that these seemingly very disparate problems might be brought together. That would be with the hypothesis that mind, rather than being a very late development in the evolution of living things, restricted to organisms with the most complex nervous systems – all of which I had believed to be true – that mind instead has been there always, and that this universe is life-breeding because the pervasive presence of mind had guided it to be so." [2]

During the 1980s, B.M. Puri was introduced to the philosophy of G.W.F. Hegel through the auspicious association of His Divine Grace Srila Bhakti Rakshak Sridhara Maharaja, Founder-Acharya of Sri Chaitanya Saraswat Math in Nabadwip, West Bengal, India. The photo below shows Srila Sridhara Maharaja seated to the right of Srila

Prabhupada, being interviewed by Srila Govinda Maharaja (the Successor-Acharya of Srila Sridhara Maharaja).

Upon returning to America in the 1990s, B.M. Puri established online Hegelian philosophy discourses [3] and started Yahoo group discussions concerning the relevance of consciousness studies in modern empirical science. During the late 2000s, in cooperation with the current President-Sevaite-Acharya of Sri Chaitanya Saraswat Math, Srila Bhakti Nirmal Acharya Maharaja, B.M. Puri remotely established the Sri Chaitanya Saraswat Institute of Spiritual Culture and Science (SCSISCS) in Bangalore, Karnataka, India. In 2012, the Bhakti Vedanta Institute of Spiritual Culture and Science was established in Princeton, New Jersey, USA. The Science and Scientist international conference series was inaugurated the following year in India.

The author humbly acknowledges the affectionate guidance, teachings, and inspiration of all the above spiritual guardians whose service he aspires to achieve by their grace. Heartfelt gratitude to Joan Walton and Brian Ford for their wonderful friendship and gracious encouragement. Fond appreciation for the services and association of Sripad Bhakti Niskama Shanta Maharaja, Sripad Bhakti Vijnana Muni Maharaja, Srimati Sumangala DD, Srimati Kushum DD, Sripad Rasaraja Prabhu, and Sriman Krishna Keshava Prabhu who are the joy of my life. Remembering always the love and support of family and friends.

References

[1] Damodara, Bhaktisvarupa. *Srila Prabhupada's Vision for the Bhaktivedanta Institute.* Retrieved from: http://www.bhaktiswarupadamodara.com/spvbi.pdf

[2] Wald, George. *Cosmology of Life and Mind.* 1988.

[3] www.gwfhegel.org

Contact Us

The Princeton BVISCS is a 501(c)(3) charitable nonprofit educational organization. Kindly make any donations at www.bviscs/donation. We humbly and sincerely thank you for your interest in our institution.

www.bviscs.org
princeton@bviscs.org

Social Media

 @bviscs

 Mahaprabhu Broadcasting Channel - MBC

www.ingramcontent.com/pod-product-compliance
Lightning Source LLC
Chambersburg PA
CBHW071018080526
44587CB00015B/2423